T4-ATW-906

Hidden Messages

By Dorothy Van Woerkom
Illustrated by Lynne Cherry

Crown Publishers, Inc., New York

Also by Dorothy Van Woerkom

The Rat, the Ox, and the Zodiac

AUTHOR'S NOTE: This little-known experiment of Benjamin Franklin's was recorded in detail by the Swedish botanist Peter Kalm in Volume 1 of his two-volume *Peter Kalm's Travels in North America (1748–1751).*

 Inquiries should be addressed to Crown Publishers, Inc., One Park Avenue, New York, N.Y. 10016. Manufactured in the United States of America. Published simultaneously in Canada by General Publishing Company Limited.

10 9 8 7 6 5 4 3 2 1

The text of this book is set in 16 point Plantin. The three-color illustrations were prepared as black line drawings, with halftone overlays prepared by the artist for black, red, and yellow.

Library of Congress Cataloging in Publication Data
Van Woerkom, Dorothy. Hidden Messages Summary: Describes Ben Franklin's experiments with ant communication, Jean Henri Fabre's with moths, and the subsequent discovery by scientists of pheromones by which animals do indeed convey messages. 1. Ants—Behavior—Juvenile literature. 2. Animal communication—Juvenile literature. 3. Pheromones—Juvenile literature. 4. Insects—Behavior—Juvenile literature. [1. Ants—Habits and behavior. 2. Animal communication. 3. Pheromones. 4. Insects—Habits and behavior] I. Cherry, Lynne. II. Title. QL568.F7V36
595.7'96 78-10705 ISBN 0-517-53520-3

For John T. Van Woerkom

There once lived a man
who was always asking questions.
His name was Benjamin Franklin,
and he lived more than
two hundred years ago.

One day he saw a dead fly
on his windowsill.
A tiny ant ran around
and around it.
The ant tried to push it.
The ant tried to pull it.
But the fly was too big.
So the ant gave up and went away.
Ben went away, too.

In a little while Ben came back.
The dead fly was still on the windowsill.
But now lots of ants were running around it.

Some of the ants picked up the fly
and carried it away. When Ben saw that,
he wondered how the other ants had
found out about the dead fly.

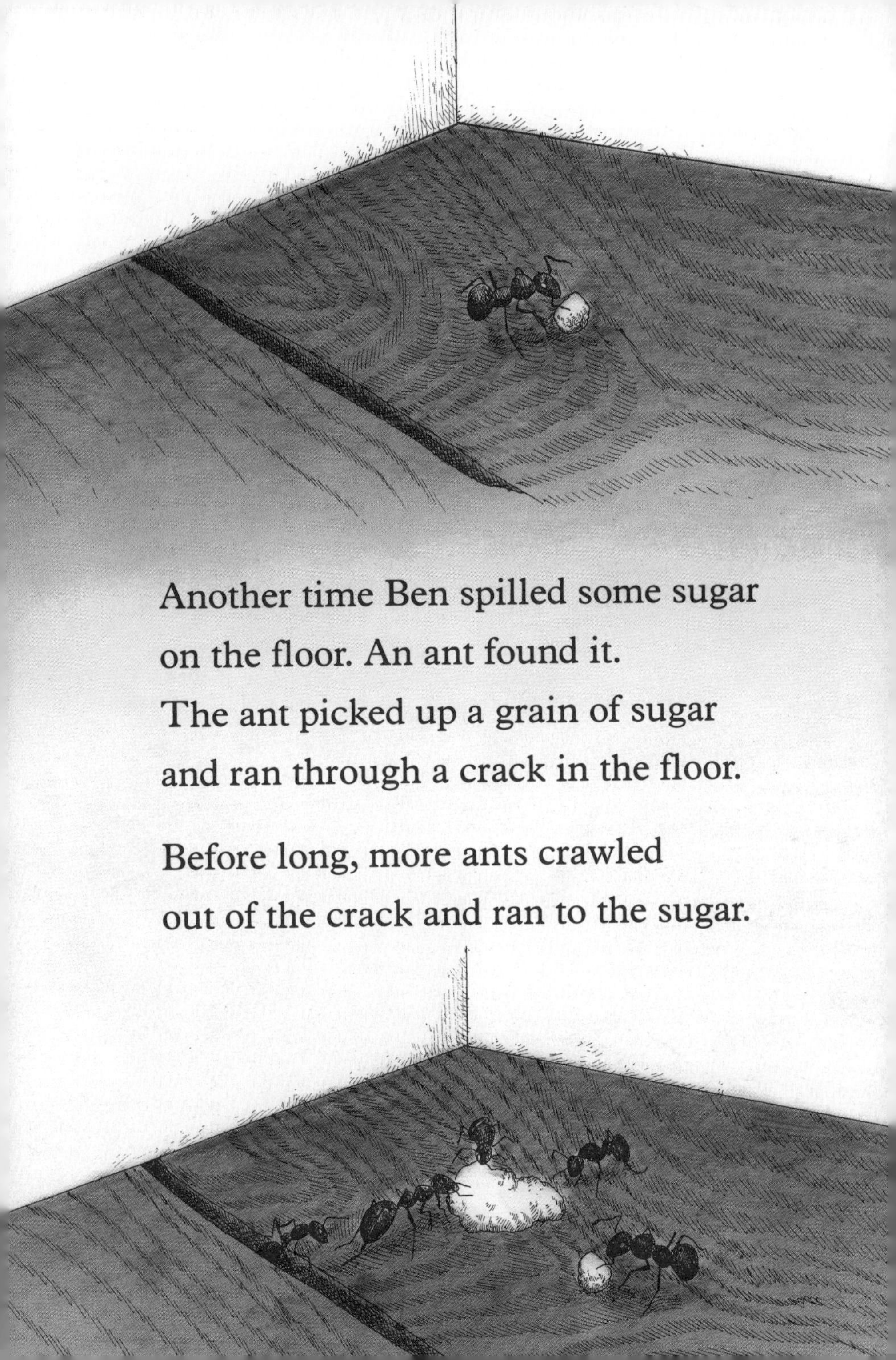

Another time Ben spilled some sugar
on the floor. An ant found it.
The ant picked up a grain of sugar
and ran through a crack in the floor.

Before long, more ants crawled
out of the crack and ran to the sugar.

Each ant took some sugar
and carried it away.
The ants kept coming until
it was all gone.

When one ant finds food,
how do the other ants
know about it?
wondered Ben.

One morning Ben wanted some molasses.
He looked into the molasses pot.
It was full of ants!

Ben turned the pot upside down
and shook it, hard.
The ants fell out.
This gave Ben an idea.
If he put the pot in another place,
would the ants find it again?

Ben drove a nail into the ceiling
and tied a string around it.
He tied the other end
to the handle of the pot.
It swung back and forth.

But Ben saw that one ant
was still in the pot.
This gave him another idea.
If he left the ant in the pot,
would it find its way back to its nest?
And if it did, would the other ants
find their way to the molasses?
Ben waited to see.

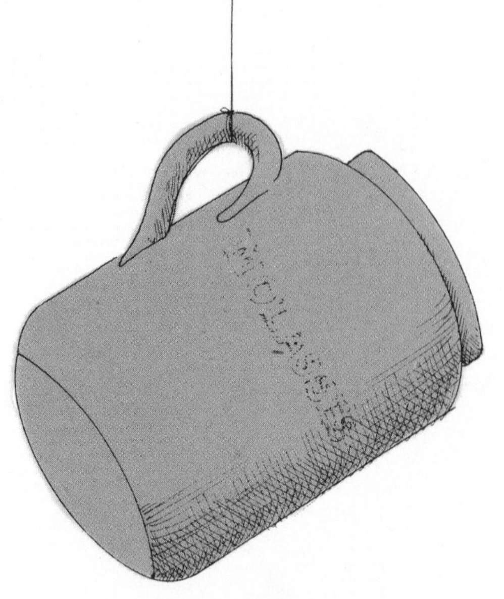

The ant ran up and down
and around the pot,
inside and outside.
Finally it found the string.
It climbed all the way up the string
and across the ceiling.
It crawled down the wall
and through the crack in the floor.

The big clock on the wall ticked away.
Five minutes . . . ten minutes . . .
fifteen minutes . . .
almost half an hour went by.

At last one ant crawled
out of the crack in the floor.
Then another, and another.

A long line of ants
crawled up the wall
and across the ceiling.
They climbed down the string
and into the pot.

Then they crawled back up the string.
They crawled across the ceiling,
down the wall, and through the crack,
just as the first ant had done.
Ants kept coming and going
until all the molasses was gone.

Now Ben was sure that ants
could give each other messages!
When one ant found something to eat,
it had some way of letting
the other ants know about it.

Ben was not the only one asking questions.
Other people found out that many kinds
of insects can give messages, but
no one knew how.

More than a hundred years later
this was still a mystery.
Then a scientist named Jean Henri Fabre
began to study moths.
One day he put a female moth into
a small cage and left his window open.
That night, dozens of male moths flew
into the room. They fluttered around
the cage, touching it with their feelers.

Every day Fabre moved the female moth to another place. Every night the male moths found her and touched her cage with their feelers.

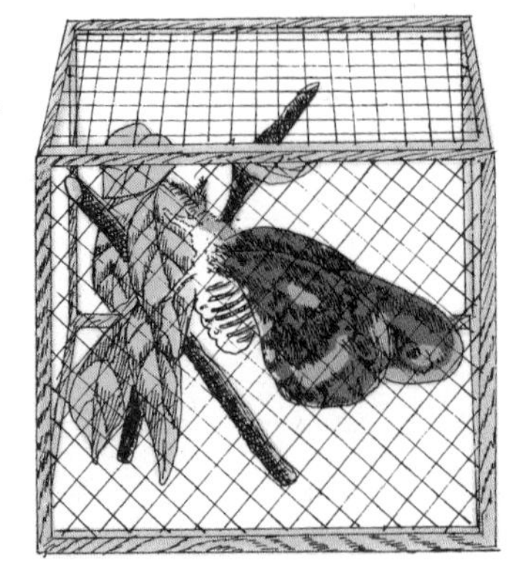

Then Fabre put the moth under a glass jar. That night, not one male moth came to visit.

When other scientists heard about
Fabre's experiment, they wondered if
the male moths could smell
the female moth with their feelers.
Perhaps that glass jar kept her odor
inside, so the moths outside
could not smell it!

Since Fabre's experiment with moths,
many people have studied insects.
Scientists now know that insects give
each other messages through odors.
These odors are called
pheromones (FER-ah-monz).
They come from a little tube
called a gland,
which is inside the insect.

An ant's gland is near
the tail end of its body.
Ants use pheromones to
mark a trail to food,
to warn the others when
an enemy enters the nest,

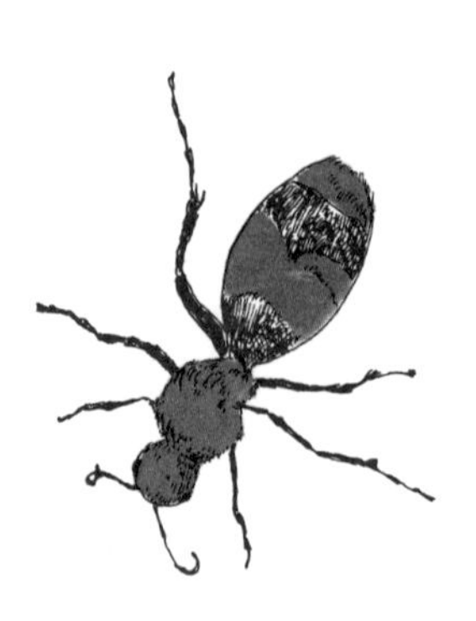

to attract a mate,
to let each other know
when to start a new nest,
and to let others know that
an ant is dead and must be
carried out of the nest.

Some ants have many pheromones.
Others have only a few.
Each pheromone has a different odor,
but they all come from the same gland.

There are almost 10,000
different kinds of ants,
and each kind has its own odors.
An ant pays attention to the odor messages
of ants from its own nest.

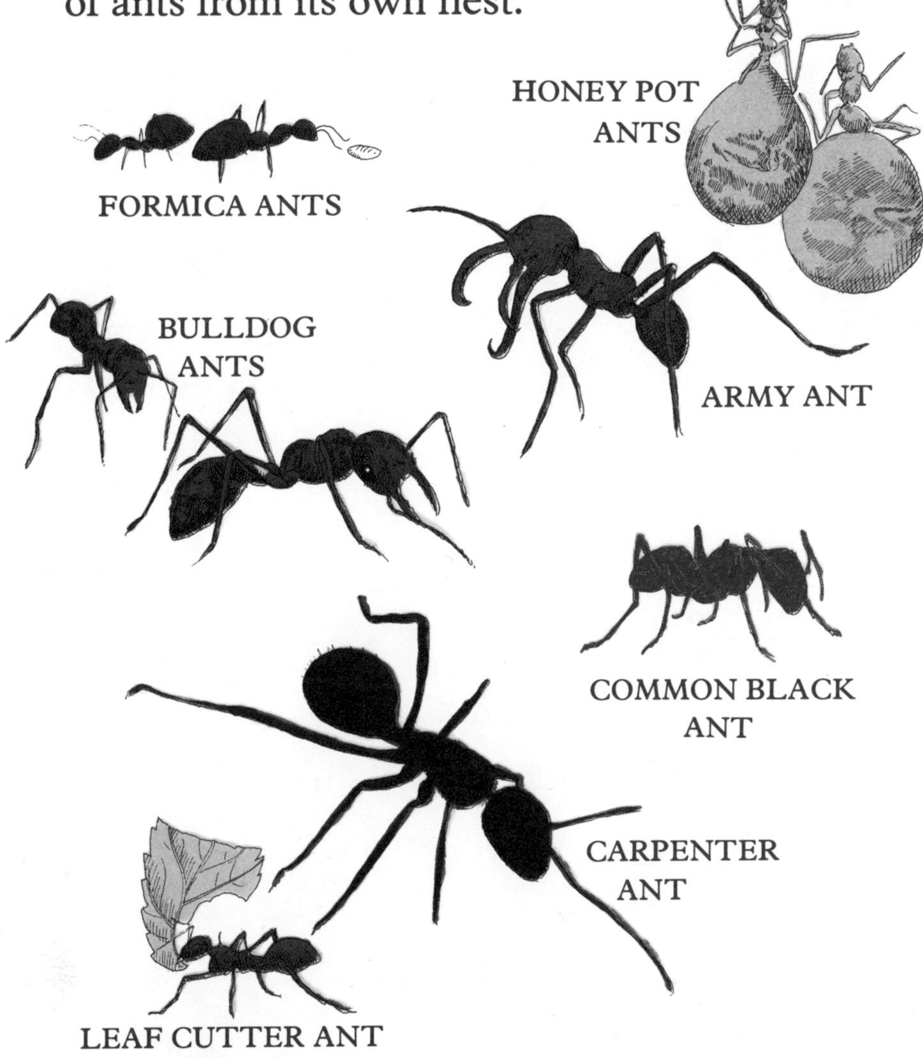

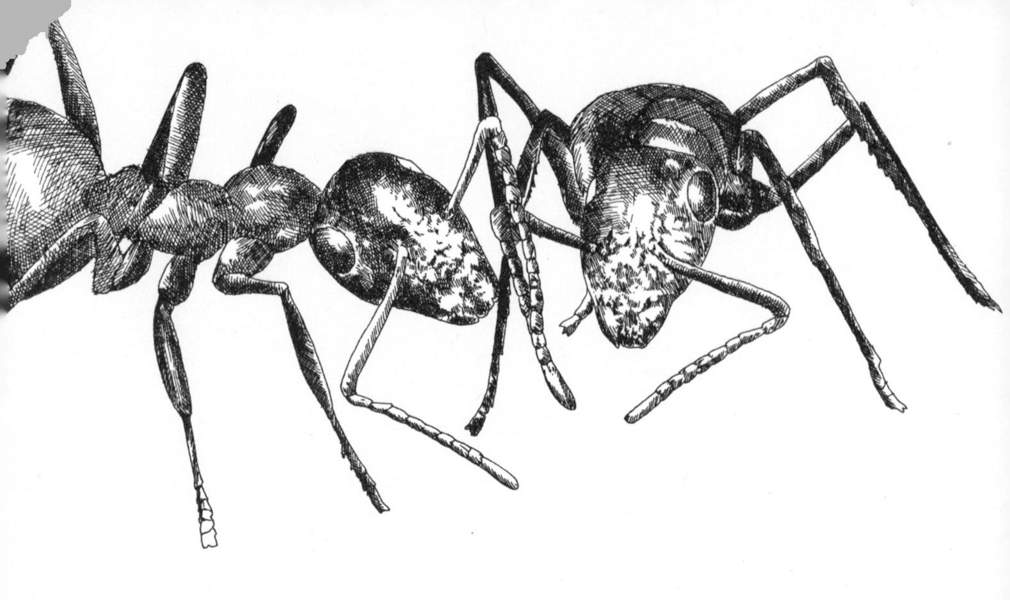

Benjamin Franklin's ants may have been
one of several kinds of common black
or red ants that we sometimes find
in our houses.
When one of these ants
finds food that it cannot carry,
it makes an odor trail back to the nest.
You would not be able to see or smell
the ant's trail, but ants can smell it.
Ants do not have noses as we do;
they smell with their feelers.

When an ant makes an odor trail,
its gland squirts out a liquid
which has an odor.
Most common household ants do this
in one of two ways.
Some ants stop many times
on the way back to their nest.
Each time they stop,
they press their bodies to the ground
and squirt out droplets of liquid.
Other ants drag their bodies along
the ground and squirt thin lines
of liquid behind them.

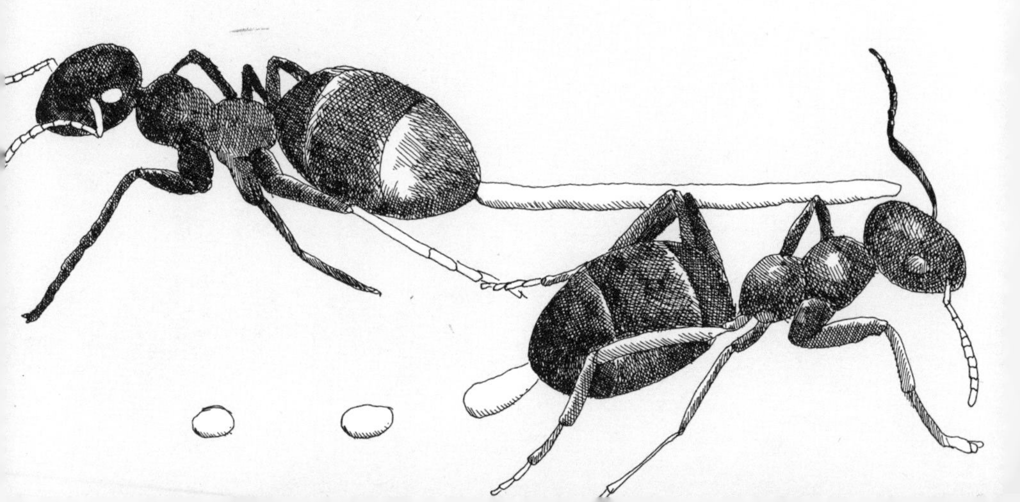

Whichever way an ant makes its trail,
some of the odor sticks
to hair on its legs.
When the ant returns to the nest,
the other ants smell it
with their feelers.
The odor gets them excited
and they leave the nest.
They use their feelers to find the
odor trail and follow it to the food.

When they return to the nest,
they leave their own trail.
More ants smell the odor on their legs,
and they too leave the nest.
But when the food is gone,
they do not make a trail,
and there is no odor on their legs
to excite the other ants
and make them leave the nest.

All ants live in groups called colonies.
Each colony has a queen and many workers.
Sometimes a colony of ants needs a new nest.
The old nest may be damaged or the ants
may no longer be able to find food nearby.

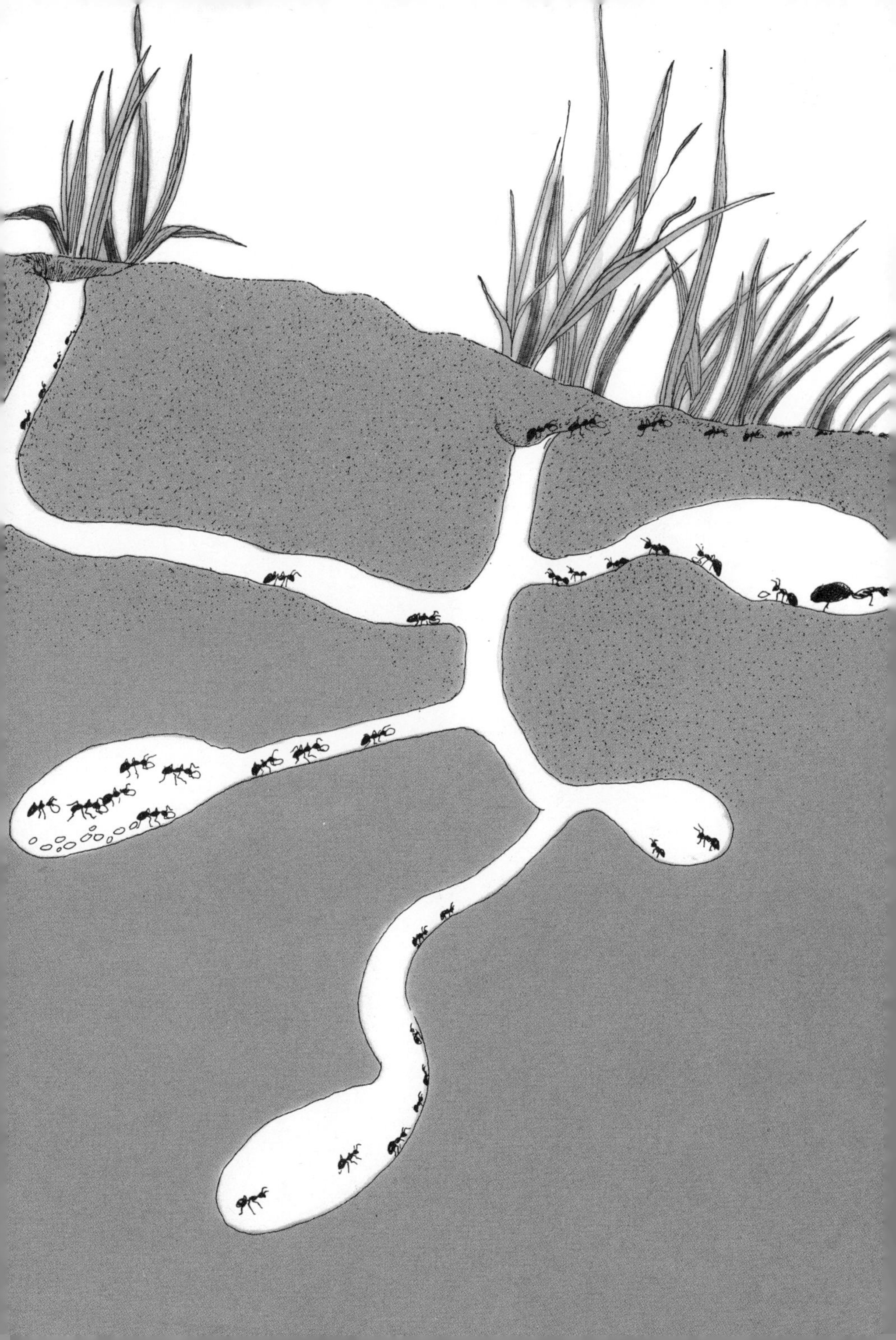

Fire ants, which live in
the southern part of the United States,
move often. When a worker
finds a good place for a new nest,
it gets excited and makes an
odor trail back to the old nest.
Other workers follow the trail
to the new place.

If they get excited too,
they make their own trail
on top of the first one.
After a while the trail has
such a strong odor that
the rest of the fire ants leave
the old nest and follow the trail
to the new place.

These are some of the many ways
that ants use odor messages.
But other insects, many larger animals,
and even fish have pheromones, too.

Do people have pheromones?
Some scientists think we do,
but they are still trying
to find out more about them.
When we want to tell each other
something, we can say it or write it.
But each person has a special odor,
different from the odor of anyone else.
We know this because dogs can find us
by following our odor.

Scientists are not sure that
this is a pheromone odor.
But they believe we may have
some pheromones.
If we do have them, can we learn
how to use them?
If we keep asking questions,
like Ben Franklin and Jean Henri Fabre,
someday we may find the answers.

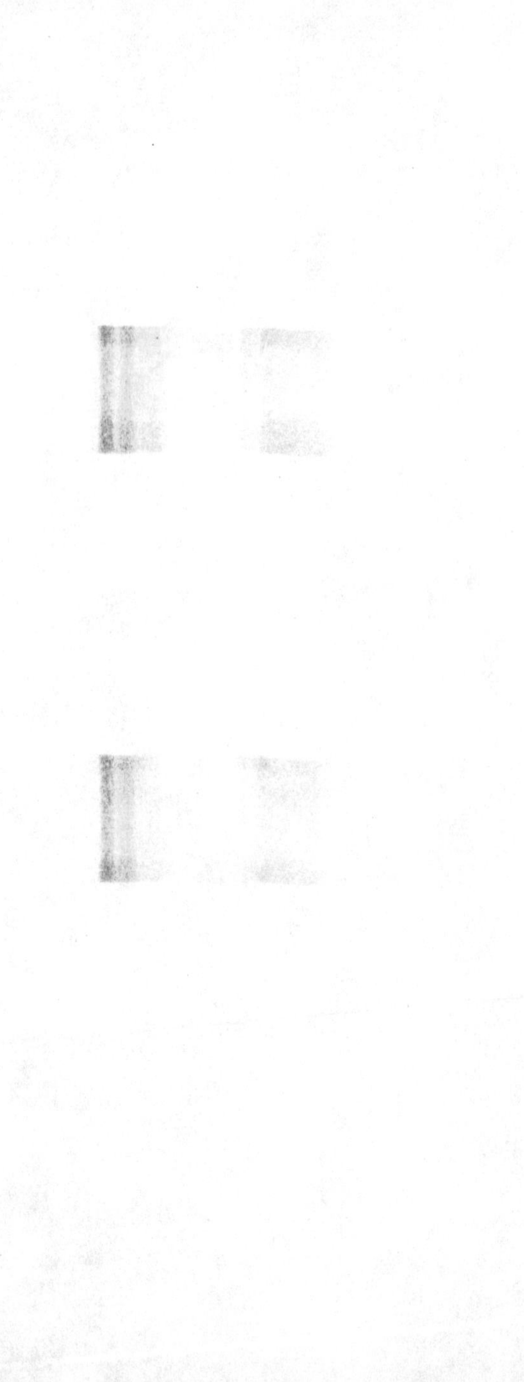